Nine Fatality Apartment House Fire
Ludington, Michigan

Investigated by: Randolph E. Kirby

This is Report 072 of the Major Fires Investigation Project conducted
by TriData Corporation under contract EMW-90-C-3338 to the United
States Fire Administration, Federal Emergency Management Agency.

FEMA

Department of Homeland Security
United States Fire Administration
National Fire Data Center

U.S. Fire Administration Fire Investigations Program

The U.S. Fire Administration develops reports on selected major fires throughout the country. The fires usually involve multiple deaths or a large loss of property. But the primary criterion for deciding to do a report is whether it will result in significant "lessons learned." In some cases these lessons bring to light new knowledge about fire--the effect of building construction or contents, human behavior in fire, etc. In other cases, the lessons are not new but are serious enough to highlight once again, with yet another fire tragedy report. In some cases, special reports are developed to discuss events, drills, or new technologies which are of interest to the fire service.

The reports are sent to fire magazines and are distributed at National and Regional fire meetings. The International Association of Fire Chiefs assists the USFA in disseminating the findings throughout the fire service. On a continuing basis the reports are available on request from the USFA; announcements of their availability are published widely in fire journals and newsletters.

This body of work provides detailed information on the nature of the fire problem for policymakers who must decide on allocations of resources between fire and other pressing problems, and within the fire service to improve codes and code enforcement, training, public fire education, building technology, and other related areas.

The Fire Administration, which has no regulatory authority, sends an experienced fire investigator into a community after a major incident only after having conferred with the local fire authorities to insure that the assistance and presence of the USFA would be supportive and would in no way interfere with any review of the incident they are themselves conducting. The intent is not to arrive during the event or even immediately after, but rather after the dust settles, so that a complete and objective review of all the important aspects of the incident can be made. Local authorities review the USFA's report while it is in draft. The USFA investigator or team is available to local authorities should they wish to request technical assistance for their own investigation.

This report and its recommendations were developed by USFA staff and by TriData Corporation, Arlington, Virginia, its staff and consultants, who are under contract to assist the USFA in carrying out the Fire Reports Program.

The USFA greatly appreciates the cooperation received from Fire Chief Mike MacDonald, Ludington Fire Department; Police Chief Walter Taranko; and Fire Investigator Detective/Sergeant Joel Dekraker of the Michigan Department of State Police

For additional copies of this report write to the U.S. Fire Administration, 16825 South Seton Avenue, Emmitsburg, Maryland 21727. The report is available on the USFA Web site at http://www.usfa.dhs.gov/

U.S. Fire Administration

Mission Statement

As an entity of the Department of Homeland Security, the mission of the USFA is to reduce life and economic losses due to fire and related emergencies, through leadership, advocacy, coordination, and support. We serve the Nation independently, in coordination with other Federal agencies, and in partnership with fire protection and emergency service communities. With a commitment to excellence, we provide public education, training, technology, and data initiatives.

TABLE OF CONTENTS

OVERVIEW 1

SUMMARY OF KEY ISSUES. 2

BUILDING HISTORY AND OCCUPANCY. 2

BUILDING CONSTRUCTION. 3

BUILDING FIRE PROTECTION 3

THE FIRE. 4

FATALITIES 5

INJURIES. 6

RESCUE AND SURVIVORS 6

CODES 6

ORIGIN AND SPREAD OF FIRE AND SMOKE. 7

DAMAGE ASSESSMENT. 7

FIRE DEPARTMENT AND EMERGENCY SERVICES 8

TENANT OVERCROWDING. 8

LESSONS LEARNED. 8

APPENDIX A: Regional Map. 11

APPENDIX B: Building Floor Plans Showing Area of Origin and Location of Fatalities 12

APPENDIX C: Fireground Diagram Showing the Placement of Units. 14

APPENDIX D: Fire Department Response Times and Units and Personnel Used at the Fire . . 15

APPENDIX E: Photographs. 16

Nine Fatality Apartment House Fire
Ludington, Michigan
February 28, 1993

Local Contact: Mike MacDonald, Fire Chief
Ludington Fire Department
201 South Williams Street
Ludington, MI 49431

Walter Taranko, Police Chief
Ludington Police Department
201 W. Loomis Street
Ludington, MI 49431

Detective/Sergeant Joel DeKraker
Fire Investigator
Michigan Department of State Police
588 Three Mile Road, NW
Grand Rapids, MI 49504
(616) 784-4996

OVERVIEW

On Saturday, February 28, 1993, at 12:18 a.m., an alarm was received by the Mason County Central Dispatch for a house fire at 208 North James Street in the city of Ludington, Michigan.

The first to arrive on the scene at 12:22 a.m. was Ludington pumper 152 from the station located approximately three blocks away. This unit was confronted with heavy smoke and fire from the second floor and first floor entrance. The Ludington Fire Chief, en route to the scene from approximately half mile away, ordered additional equipment for manpower purposes from neighboring Pere Marquette Township located approximately two miles away. The fire was knocked down in about 20 minutes, and the scene was declared under control in less than one hour.

1

SUMMARY OF KEY ISSUES

Issues	Comments
Cause	Undetermined. Believed to have originated at or near a wall-mounted light fixture in second floor corridor.
Firefighting	Heavy smoke and fire throughout the second floor corridor and first floor entrance made entry difficult. Rapid response and fire extinguishment prevented total roof collapse.
Smoke and Flame Spread	Heavy smoke and flame spread due to highly combustible wall paneling and ceiling tile in second floor corridor. Air transfer grills between apartments and corridor provided additional path for smoke and heat penetration.
Means of Egress	Two first floor apartments serviced by a corridor with one means of egress transgressing an open and unprotected stairwell servicing the second level corridor which provided the sole means of egress for two additional apartments.
Smoke Detectors	Battery operated smoke detectors were improperly installed within the apartments and were ineffective.
Fire Protection Equipment	No alarm system, no exit lights or signs, questionable egress lighting, non-fire rated corridor protection, no smoke detection capability in corridor. Building was not sprinklered.

Rapid smoke and flame spread throughout the second floor, claiming the lives of nine occupants and injuring one.

The fire department used 44 firefighters and police personnel and eight units. The fire was confined to the first floor entrance foyer and second floor area.

BUILDING HISTORY AND OCCUPANCY

The apartment building is located in the city of Ludington in a residential neighborhood. It is adjoined on all sides by other residential properties. It is two stories in height with a half basement and was constructed around 1882 as a single residence.

In the early 1900s, the building was converted to hotel use and continued as such for an unknown period of time. At some point, it was changed from a hotel to a small apartment building and was divided into five one-bedroom units, two on the upper level and three on the main level.

Only two complaints had been received by Ludington building officials in the past 12 years, and they dealt with trash accumulation in the backyard and a dilapidated garage.

Previous fire experience indicates that there was a fire some 12 years ago in a first floor unit due to careless smoking. There was one fatality.

The building has changed ownership several times since its construction and was acquired by its present owners in May 1992. The five apartments were rented as follows:

- Second floor north apartment was occupied by four people.

- Second floor south apartment was occupied by seven people.

- Each of the three units on the first floor was occupied by one person.

The building appeared to be reasonably well maintained.

BUILDING CONSTRUCTION

This building is a concrete block facility, two stories high with a half basement. The floor joists are of 2 x 8-inch wood construction with a wood plank subfloor. The walls are of lath and plaster construction. The hip roof is constructed of wood trusses with asphalt shingles. The overall dimensions of the building are 30 feet 4 inches by 53 feet 6 inches. It is located on a lot 60- by 140-foot.

The building is equipped with 50-amp electrical service. There are no records as to when this was installed or inspected.

Each apartment is heated by a through-the-wall, gas-fired, forced-air heater and each has several windows approximately 24 by 60 inches with a sill height below 40 inches. The windows are operable from the inside without the use of tools and are located within 20 feet of grade.

The upper level is serviced by a corridor approximately 44 inches wide. The north end corridor is open to an unenclosed stairwell which opens to the lower level. The east end of the upper corridor leads to a fire escape which is accessible through an interior stairwell approximately one-half level down. This stair is enclosed at the top by a wood panel door with a self-closure. There is no fire-resistant protection from the floor below. The exterior fire escape is not protected from the elements and is not of fire-resistant construction.

The two front lower level apartments are serviced by a corridor with one means of egress transgressing the open stairwell servicing the second level. In addition, these apartments have a second means of egress directly to the outside. The rear apartment is not served by this corridor and has direct egress to the outside.

Thirty-two-inch wood panel doors are used throughout the building. The interior finish consists of non-rated pre-finished wood paneling approximately 3/8-inch thick, installed over lath and plaster walls in the corridors and 12- by 12-inch wood fiber ceiling tiles attached to 1- by 3-inch wood furring strips installed on the ceiling throughout the building. Some wood paneling was used in the individual apartment units, as well as paper and vinyl wall coverings.

The building is equipped with louver openings approximately 13 by 9 inches. They are located between the living units and corridors at various points and are wall-mounted just below the ceiling. The louvers were in the closed position. They had been covered over with wood paneling on the corridor side and with wallpaper on the room side.

BUILDING FIRE PROTECTION

There was no alarm system or smoke detection equipment installed in the corridors. There was no emergency lighting, such as a battery pack. There was no sprinkler system.

Each apartment unit contained one battery-operated smoke detector which had been improperly installed on the wall approximately five feet off the floor. It is believed that some, if not all, were operable at the time the fire department arrived on the scene. However, because of their incorrect location, they failed to render an early warning to the occupants.

Each unit was equipped with a 1A-10BC fire extinguisher. These were mounted by a bracket on the kitchen walls.

The single means of egress servicing the second floor was through an unrated, unprotected corridor.

THE FIRE

On February 28, 1993, just after midnight, the male occupant of a second floor apartment was awakened in his bedroom by his wife who apprised him of a fire situation in their apartment. He immediately left the bedroom and entered the living room where his wife and children had been sleeping. He noticed fire and heavy smoke coming through the closed corridor vents installed in the wall between his living room and the corridor. He went into his kitchen and got a 10-pound dry chemical fire extinguisher, returned to the living room, where he grabbed his 3-year-old son and went out the apartment door into the corridor. He noticed fire and smoke head high in the vicinity of a light fixture mounted on the south wall of the corridor. He then released his son's hand in order to pull the pin on the fire extinguisher. He discharged the extinguisher in the direction of the flames.

Realizing his effort was ineffective, he turned to leave and noticed that his son was not with him. He re-entered his apartment to get his family members, but was unable to find them. He went back into the hall and noticed fire and smoke traveling east along the wood paneled walls. At this time, burning ceiling tile began to fall on his unclothed body. While trying to protect himself, he fell down the steps to the first floor and exited the building. He began yelling for help and knocking on the first floor apartment doors. Unable to summon anyone, he ran to the rear of the building and knocked on the door. The occupant of this unit called the telephone operator who, in turn, notified the fire department of the incident.

Notification of the alarm was received by central dispatch at 12:18 a.m. At 12:19, units were dispatched. At this same time, a Ludington police officer, who was approximately one-half block away, arrived on the scene moments after the units were dispatched. He reported large clouds of smoke emitting from the rooftop and from the entire second floor. He also observed the occupant of the second floor north apartment trying to re-enter the building through the front entrance. The officer approached the man and tried to obtain information regarding the other occupants, but, because of his state of anxiety, conflicting answers were given. The officer then returned to his vehicle where he donned self-contained breathing apparatus. He went to the rear of the building to see if he could enter by way of a wooden, exposed fire escape, but flames and smoke prevented its use.

At 12:22 a.m., Unit 152 of the Ludington Fire Department arrived on the scene with two firefighters. They laid a 5-inch supply line from a hydrant some 300 feet away to the front of the building where a handline was advanced to the front door. At 12:25, Unit 155 arrived with three personnel and they positioned themselves in front of the building. The firefighters placed two handlines in service – one to the rear of the property and the second to the south side front of the building. At 12:26, Unit 154 with four firefighters arrived on the scene. The initial attack on the fire was made through the front entrance of the building and from the south side of the first floor. Handlines were also advanced up the rear fire escape to gain entrance to the rear portion of the building.

At approximately 12:26 a.m., the Ludington Fire Chiefs while en route to the scene, requested additional units from Pere Marquette, a neighboring township, for manpower purposes. At 12:33, Unit 157 (Ludington) arrived on the scene with three personnel.

At 12:35 a.m., Pere Marquette Unit 29-2 arrived and positioned itself in the rear of the building where a handline was placed into service up the rear fire escape.

During the next few minutes, three additional units from Pere Marquette arrived on the scene, and their manpower was used to augment the overall firefighting effort. At approximately 12:41 a.m., the major portion of the fire had been knocked down.

Firefighters were able to gain access to all portions of the building at this time, where they discovered three bodies in the second floor north apartment bedroom and six bodies in the second floor south apartment living room. The victims in the south apartment appeared to have been asleep on the floor and never woke up. One boy who did not live in the south apartment apparently went into that apartment while his father was fighting the fire in the hallway. All the bodies were removed and transported to the morgue. During the overhaul operation, a melted battery-operated smoke detector was found on the floor in the living room by the north wall of the apartment where the six fatalities occurred. Above it was an air transfer vent which allowed smoke and heat into this room from the corridor. It appears that this detector was mounted somewhere below this opening.

Fire damage revealed that the fire originated in the corridor at or below the ceiling level in the vicinity of a light fixture on the south wall. Fire progressed rapidly because of the highly combustible wood paneling and wood fiber ceiling tiles. It was aided by an additional supply of oxygen as a result of the occupant leaving both his apartment door and the front door open when exiting the building. It was allowed to penetrate the north and south apartments at an early stage, burning through wooden doors and through air-transfer grills that were installed in the corridor walls that separated the living units.

The scene was declared under control at 12:55 a.m. The services of 44 fire, police, and ambulance personnel were employed.

FATALITIES

There were nine fatalities. All occurred in the two apartments on the second floor and all but one, the babysitter, were residents of the building. Three bodies were found in the north apartment bedroom and six were found in the south apartment living room.

North Apartment

> One-year-old male
> Two-year-old male
> Eighteen-year-old female

South Apartment

> One-year-old male
> Two-year-old male
> Three-year-old male (from north apartment)
> Three-year-old female
> Six-year-old female
> Thirteen-year-old female (babysitter)

The positions of the victims located in the south apartment indicate, with one exception, that all were asleep on the living room floor and never woke up. Some were still clutching stuffed animals or other objects children normally sleep with.

The three victims in the north apartment, along with the 3-year-old male, who, at some point, went into the south apartment, were aware of the fire, as the 18-year-old mother was the person who first discovered the fire. She and two of her children went into the bedroom of their north apartment, probably to shield themselves from the fire, and appeared to have been overcome by smoke before deciding to attempt escape. The body of her 3-year-old boy was found in the south apartment living room.

Toxicology reports on the victims indicated high levels of carbon monoxide – in excess of 65 percent. None of the victims appeared to have been subjected to actual flames (see Exhibits G and H.)

Officials were surprised to find so many young children in a one-bedroom apartment. It was later discovered that two families resided in the south apartment where the majority of fatalities occurred.

INJURIES

There was one injury. The sole surviving occupant of the second floor north apartment sustained 25 percent burns on his back, shoulders, and neck due to burning ceiling tile falling on him in the corridor. He was transported to the Ludington Memorial Hospital where he was admitted.

RESCUE AND SURVIVORS

It is almost certain that the fatalities in this incident had already occurred by the time the first fire-fighting equipment arrived. The advanced stages of the fire and smoke precluded the possibility of survival and posed a tremendous obstacle to entering the building, making rescue attempts impossible. However, at the point where the male occupant of the north apartment was made aware of the fire by his wife, had he led his family out of the building rather than attempting to extinguish the fire, his family may have survived.

Surviving the children found in the south apartment were the two mothers who were absent from home the night of the fire. They had engaged the 13-year-old babysitter to care for the children.

In addition, the two occupants of the two first floor front apartments were not home at the time of the fire.

The occupant of the first floor rear apartment was made aware of the fire by the occupant of the second floor north apartment. She survived.

CODES

The city of Ludington currently uses the 1990 edition of the BOCA Building Code. It is the opinion of the building official that this building did not come under the jurisdiction of current codes as it was constructed prior to the adoption of the code which exempted it from present day requirements.

In 1991, the Ludington City Commission held public hearings on a proposed rental inspection ordinance that would have made inspection of all city rental units mandatory. Most of the people present at this hearing were landlords and voiced strong objection to the ordinance. The ordinance never got beyond the committee stage.

The city fire department does attempt to make some annual inspections and familiarization trips to various commercial and industrial occupancies throughout the city. It uses as its reference the 1987 BOCA Fire Prevention Code. This code has not been adopted by the city commission and does not have the force of law.

There is virtually no inspection program for existing buildings now in place. The current building code enforcement responsibility is conducted by one person. New construction inspections and some complaints leave little or no time to take on added responsibilities.

City records indicate that the last inspection of the James Street property occurred in 1981 following a fire. The building inspector also recalls only two complaints in recent years, one concerning excess garbage and the other a dilapidated garage which was demolished in 1991.

The Michigan State Fire Marshal's Office indicates that, if this building were subject to today's codes, the following basic upgrades would have been mandatory:

- The means of egress could not be exposed to unprotected vertical openings.

- Interior stairways would be enclosed with 20-minute fire barriers.

- Exterior stairs would be reasonably protected against blockage by fire.

- Interior finish would be a minimum of Class A or B in the means of egress areas.

- At least one manual fire alarm station would be provided to initiate a fire alarm.

- Apartments would be separated from a corridor by smoke-resistant walls and self-closing doors.

- At least one fire extinguisher would be provided at the stairwell landing.

- Smoke detectors would be installed in accordance with NFPA pamphlet #74.

- Fire exit drills would be required at sufficient frequencies to familiarize all occupants as to how to exit a building safely.

ORIGIN AND SPREAD OF FIRE AND SMOKE

The fire is believed to have been of accidental nature, originating in or around a second floor corridor wall light fixture located in the vicinity of the entrance door to the south apartment. Burn patterns reveal the source of the fire originating at this location and moving east down the corridor, igniting plywood paneling on the walls and combustible ceiling tiles. Fire developed very rapidly due to the presence of sufficient fuel and the lack of early detection. It quickly penetrated the upper ceiling joist area and gained entrance into the two second floor apartments by way of the wood panel doors and in and around air-transfer vents that were exposed once the paneling had burned away.

The spread of smoke and heat was exacerbated when the second floor occupant attempted to extinguish the fire with a portable fire extinguisher, leaving his apartment door open, which provided additional oxygen as well as another avenue for smoke and fire to travel. It is believed that his attempt to extinguish the blaze did not delay notification to any appreciable degree; however, it probably negated his one and only opportunity to evacuate other family members of the north apartment.

A neighbor who lives in a house to the rear of the apartment building stated that she went by the building at about 12:10 a.m. and detected nothing unusual. This, again, indicates that the fire developed and spread rapidly.

DAMAGE ASSESSMENT

Fire was primarily confined to the first floor entrance foyer, the stairway, and the second floor corridor and apartments. There was heavy fire damage to the corridor ceiling joists in the attic area above the suspected area of origin and to the two apartments on the second floor where fire gained entry through doors and above the ceiling where fire penetrated the attic area.

There was fire penetration to the two first floor front apartments after burning through the entrance doors. The living rooms of each were damaged. The remaining rooms of the units sustained light water and smoke damage. The rear apartment suffered minor smoke damage.

The relatively light damage this building suffered in spite of such a rapidly moving fire is due solely to the quick response, excellent size-up, and effective fire suppression techniques employed, which enabled the fire department to suppress the fire in a remarkably short period of time.

Estimated damage to the building is $50,000.

FIRE DEPARTMENT AND EMERGENCY SERVICES

The city of Ludington's one fire station, located in the downtown area, operates as a paid volunteer unit. It is comprised of approximately 35 firefighters and houses one squad truck, one minipumper, and three pumpers. All were used in the James Street fire.

As a supplement to the city's firefighting and rescue effort, the police department has trained its officers in the use of breathing apparatus for use in emergency situations. Each squad car is equipped with an air pack.

The Ludington Memorial Hospital operates the ambulance service for the city. Immediately upon notification of the James Street fire incident, it sent three units to stand by at the scene. American Red Cross mobilized and implemented a plan to house and assist anyone requiring aid.

TENANT OVERCROWDING

The city of Ludington does have a regulation that determines the permissible number of people who can safely and adequately reside in living units. This requirement is in the BOCA Building Code currently used in this jurisdiction.

It is believed that the two families who lived in the second floor south apartment moved in together within the last few months. According to officials, the city was unaware that two families consisting of four children and two adults were residing in a one-bedroom apartment in the James Street building. Had the officials been made aware of it, immediate action would have been taken to alleviate the situation.

LESSONS LEARNED

1. **Code officials need the authority to require fire protection upgrades in existing buildings.**

    One of the major problems confronting the fire service today is the problem of existing buildings and how they affect the fire experience of the community. Time and time again, the fire service is experiencing tragic consequences because of inadequate and/or nonexistent code authority to alleviate dangerous and hazardous conditions that exist in these buildings. Code officials should be provided with the authority to require basic minimum fire protection upgrades in older buildings where situations present a clear danger to the occupants of these structures.

    The James Street incident illustrates the urgent need for such authority. As presented elsewhere in this report, the Michigan State Police, Fire Marshal's Division, stated that, had this building been subject to current codes, basic minimum upgrades would have been mandated. Among these are noncombustible corridors and effective smoke detection devices. It is reasonable to assume that the outcome of this incident would have been different had this building been subject to current codes.

2. **Local ordinances, following nationally recognized standards, should mandate smoke detector protection in all residential occupancies.**

This incident emphasizes the urgent need for an immediate, effective smoke detector ordinance which would mandate the installation of smoke detectors in every dwelling unit not covered by existing codes. Since the James Street tragedy has occurred, this jurisdiction, along with neighboring jurisdictions, are seriously considering adopting such an ordinance.

While this is a positive sign, care should be exercised that ordinances are not hastily developed and adopted which are not consistent with nationally recognized standards. As shown in this fire, good intentions were shown in that smoke detectors were provided; unfortunately, the recommendations and instructions by national standards were either not known or not followed, which rendered this well-intended action ineffective.

3. **Effective inspections of multifamily residential occupancies are an essential part of fire protection in a community.**

By far some of the most serious hazards facing a community regarding fire are with pre-code residential buildings. Timely and thorough inspections of such properties are the most effective means of detecting and correcting hazardous conditions that affect the safety of occupants in these buildings. Such a program must have dedicated to it an adequate number of properly trained personnel. Potential resources for implementing and enforcing such a program should include not only the fire department but also local housing, health, and welfare agencies which can assist in identifying overcrowding conditions and other health and safety concerns. An annual rental inspection program is a widely used and an effective method of identifying and correcting hazards.

4. **State leadership in code authority and administration can improve fire protection in individual communities throughout the State.**

As with any tragedy, immediate solutions are on the minds of everyone. However, immediate choices are not always the proper ones. A more effective approach to uniformity in code development would be for the Michigan State legislature to adopt a uniform building code and fire code which would be mandatory in all jurisdictions throughout the State. This would eliminate doubt as to which requirements are in effect concerning certain conditions regardless of geographical location. It would also remove the problem of politics at the local level as to which codes would be adopted or rejected. A good example is the rental inspection program proposed by the Ludington Commission in 1991, but never enacted.

5. **Positive actions by State and local fire officials can help the community cope with the emotional side of a tragedy and also enhance the spread of fire prevention education.**

This fire tragedy, representing the single largest loss of life in Ludington, has had a profound effect on the community. According to the local residents and to published reports, everyone was grief-stricken, confused, and perhaps feeling a bit insecure. Compounding this state of despair were hundreds of news inquiries from all over the Nation seeking information about the tragedy.

Recognizing this situation, the city police chief and the Michigan State Fire Marshal's Office, along with other key city officials, immediately released sufficient information about the fire in order to put to rest any concerns regarding the fire cause. In addition, the quick and excellent method of investigation by local and State authorities working together to determine the origin of the fire enabled them to correct inaccurate information that was somehow filtering out.

Professional counseling services were made available for citizens and fire department personnel. Special community meetings were held where State and local authorities answered citizens' questions and provided them with fire prevention information concerning steps to take in fire situations. This kind of positive action is essential in any community that experiences such a tragic event.

Regional Map

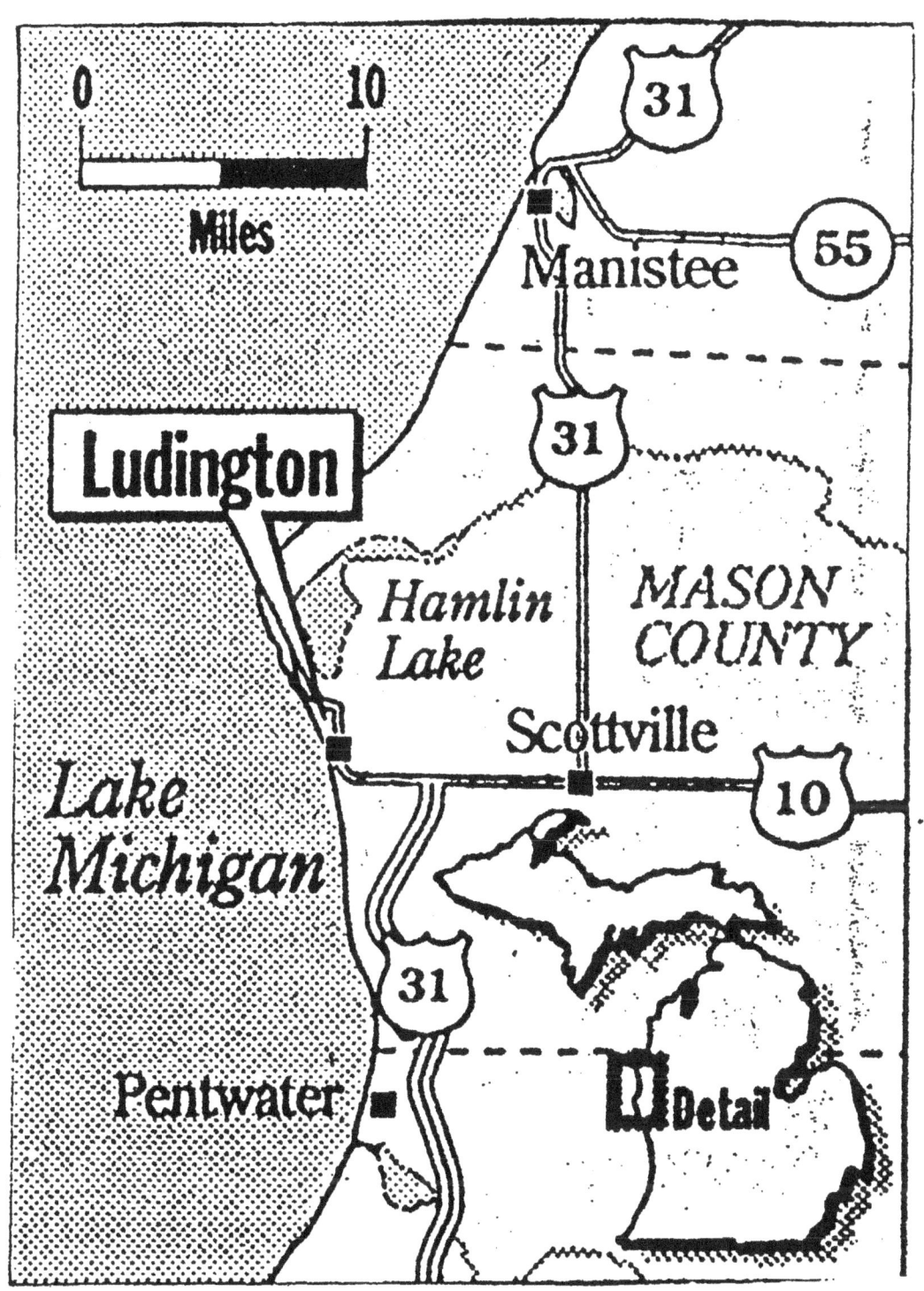

APPENDIX B

Building Floor Plans Showing Area of Origin and Location of Fatalities

FIRST FLOOR PLAN

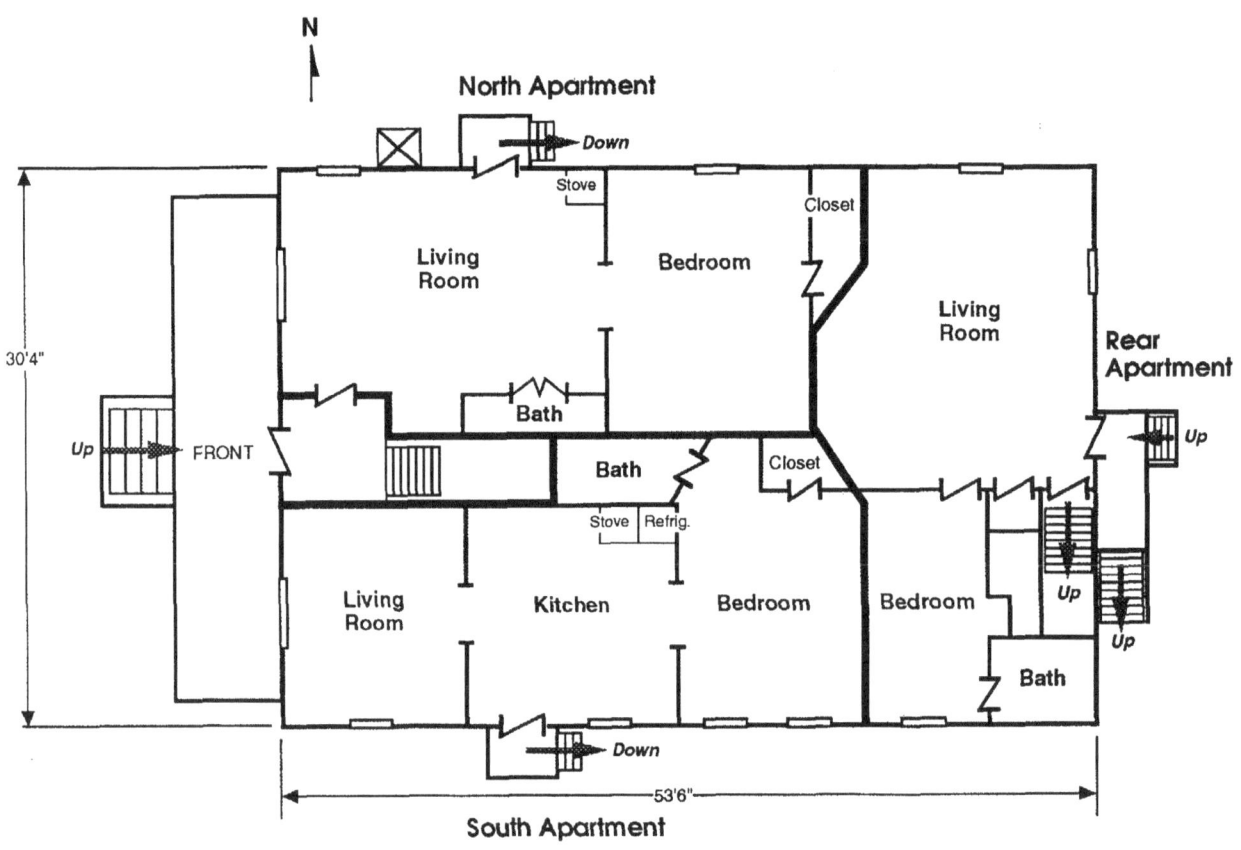

772-3-24-93-2

12

Appendix B (continued)

SECOND FLOOR PLAN

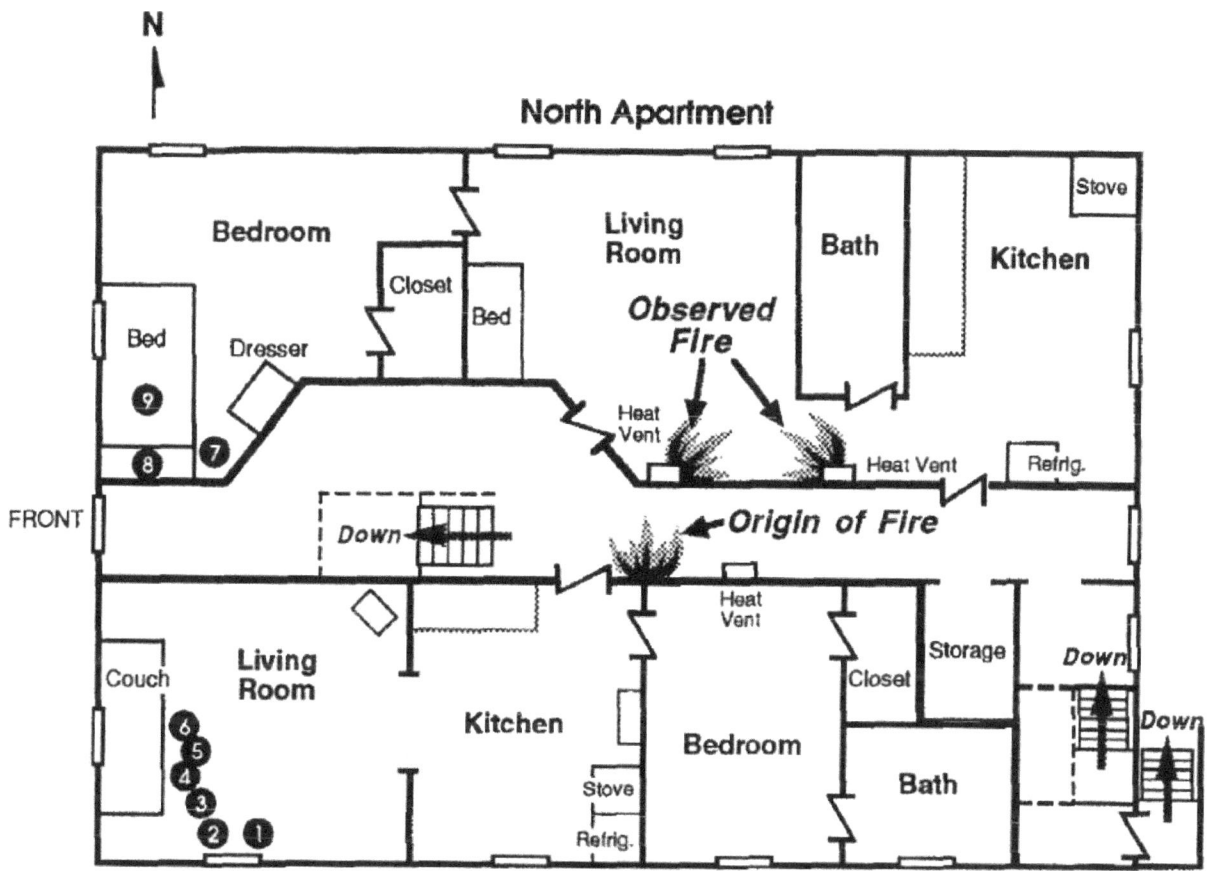

FATALITIES

1. 3-year-old girl
2. 2-year-old boy
3. 3-year-old boy
4. 1-year-old boy
5. 6-year-old boy
6. 13-year-old girl (babysitter)
7. 1-year-old
8. 18-year-old woman
9. 2-year-old boy

772-3-24-93-1

APPENDIX C

Fireground Diagram Showing the Placement of Units

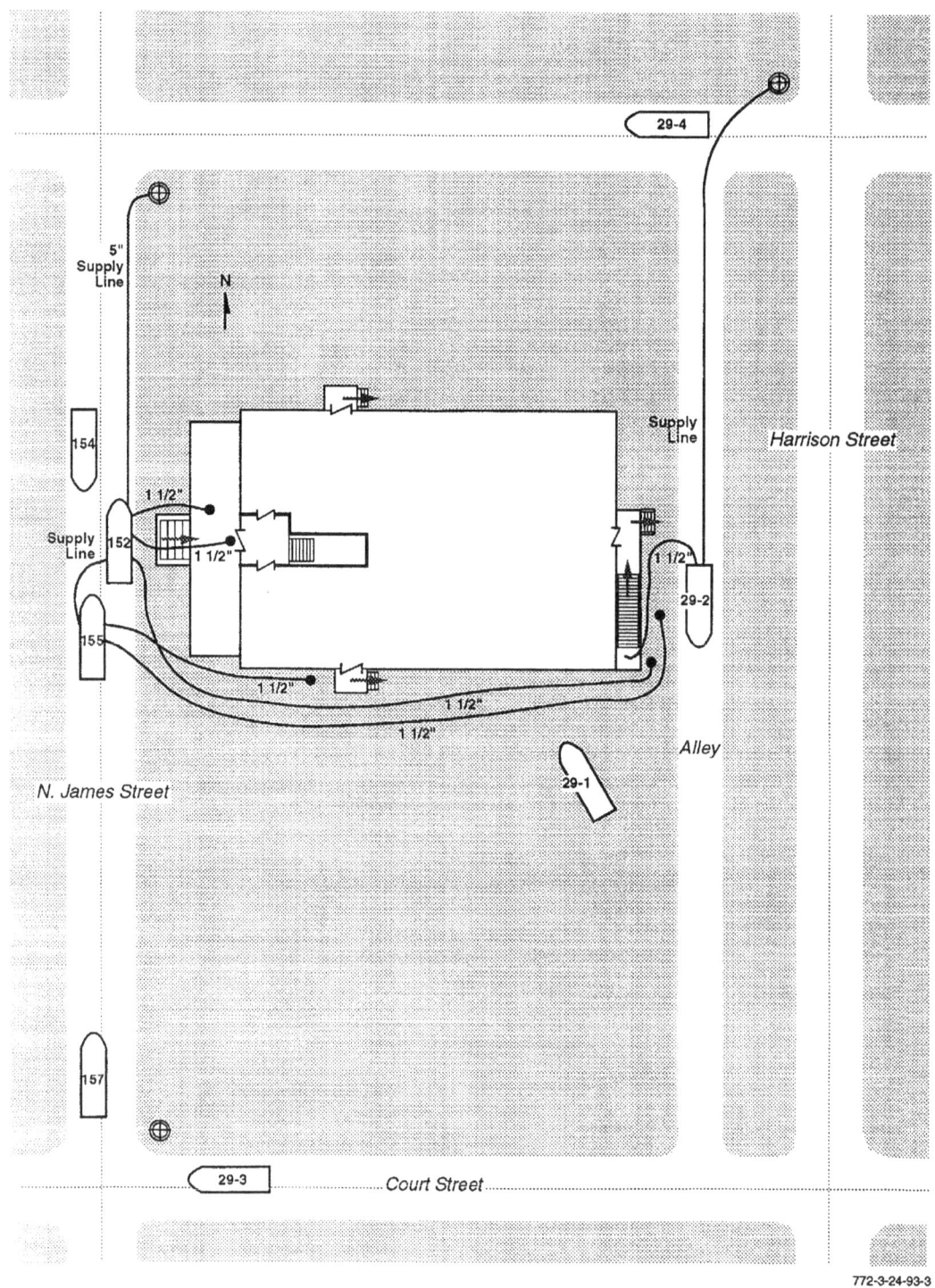

14

APPENDIX D

Fire Department Response Times and Units and Personnel Used at the Fire

FIRE DEPARTMENT RESPONSE TIMES

12:18 a.m.	Alarm Received
12:19 a.m.	Units Dispatched
12:19 a.m.	Police Officer on Scene
12:22 a.m.	Unit 152 on Scene
12:25 a.m.	Unit 155 on Scene
12:26 a.m.	Unit 154 on Scene
12:26 a.m.	Fire Chief Calls for Mutual Aid
12:33 a.m.	Unit 157 on Scene
12:35 a.m.	Unit 29-2 on Scene
12:38 a.m.	Unit 29-1 on Scene
12:40 a.m.	Unit 29-3 on Scene
12:41 a.m.	Main Body of Fire Knocked Down
12:43 a.m.	Unit 29-4 on Scene
12:55 a.m.	Scene Under Control

UNITS AND PERSONNEL USED AT THE FIRE

Ludington Fire Department

3 Engines

1 Squad Truck (Heavy Rescue)

19 Firefighters and Officers

5 Police

6 Ambulance Personnel

Pere Marquette Fire Department

3 Engines

1 Tanker

20 Firefighters

APPENDIX E

Photographs

Photo by Randolph Kirby

Front of building facing James Street.

Appendix E (continued)

Photo by Randolph Kirby

Separate entrance to first floor rear apartment and wooden fire escape to second floor.

Appendix E (continued)

Photo by Randolph Kirby

Area on second floor south apartment living room where six fatalities were found. Notice little or no actual fire damage.

Appendix E (continued)

Bed in living room of second floor north apartment; area where three fatalities were discovered. Notice heat and smoke damage only.

Photo by Randolph Kirby

Appendix E (continued)

Photo by Randolph Kirby

**Stairway from first to second floor showing burn damage to staircase,
wall, and ceiling.**

Appendix E (continued)

Photo by Randolph Kirby

Fire damage to entrance doorway and walls of the second floor north apartment
living room. Notice height of burn patterns and suspected area of origin.

Appendix E (continued)

Photo by Randolph Kirby

Smoke and heat traveled through air transfer vent from corridor to second floor south apartment in bedroom.

Appendix E (continued)

Photo by Randolph Kirby

Fire traveled through air vent from suspected area of origin into second floor north apartment living room. Again, notice high burn pattern.

Appendix E (continued)

Photo by Randolph Kirby

Fire damage to second floor corridor walls and ceiling. Notice fire penetration through wall into north apartment.